BIOCHEMISTRY RESEARCH TRENDS

RELAXATION PHENOMENA IN CHITOSAN FILMS

BIOCHEMISTRY RESEARCH TRENDS

Additional books in this series can be found on Nova's website under the Series tab.

BIOCHEMISTRY RESEARCH TRENDS

RELAXATION PHENOMENA IN CHITOSAN FILMS

EVGENY PROKHOROV
J. BETZABE GONZÁLEZ-CAMPOS
G. LUNA-BÁRCENAS
ROSA E. N. DEL RÍO-TORRES
AND
L. CHACON-GARCIA

Nova Science Publishers, Inc.
New York

Copyright © 2011 by Nova Science Publishers, Inc.

All rights reserved. No part of this book may be reproduced, stored in a retrieval system or transmitted in any form or by any means: electronic, electrostatic, magnetic, tape, mechanical photocopying, recording or otherwise without the written permission of the Publisher.

For permission to use material from this book please contact us:
Telephone 631-231-7269; Fax 631-231-8175
Web Site: http://www.novapublishers.com

NOTICE TO THE READER

The Publisher has taken reasonable care in the preparation of this book, but makes no expressed or implied warranty of any kind and assumes no responsibility for any errors or omissions. No liability is assumed for incidental or consequential damages in connection with or arising out of information contained in this book. The Publisher shall not be liable for any special, consequential, or exemplary damages resulting, in whole or in part, from the readers' use of, or reliance upon, this material. Any parts of this book based on government reports are so indicated and copyright is claimed for those parts to the extent applicable to compilations of such works.

Independent verification should be sought for any data, advice or recommendations contained in this book. In addition, no responsibility is assumed by the publisher for any injury and/or damage to persons or property arising from any methods, products, instructions, ideas or otherwise contained in this publication.

This publication is designed to provide accurate and authoritative information with regard to the subject matter covered herein. It is sold with the clear understanding that the Publisher is not engaged in rendering legal or any other professional services. If legal or any other expert assistance is required, the services of a competent person should be sought. FROM A DECLARATION OF PARTICIPANTS JOINTLY ADOPTED BY A COMMITTEE OF THE AMERICAN BAR ASSOCIATION AND A COMMITTEE OF PUBLISHERS.

Additional color graphics may be available in the e-book version of this book.

LIBRARY OF CONGRESS CATALOGING-IN-PUBLICATION DATA

Relaxation phenomena in chitosan films / Evgeny Prokhorov ... [et al.].
   p. cm.
  Includes index.
  ISBN 978-1-61761-315-9 (softcover)
  1. Chitosan. 2. Relaxation phenomena. I. Prokhorov, Evgeny.
  TP248.65.C55R45 2010
  660.6'3--dc22
                    2010031347

Published by Nova Science Publishers, Inc. † New York

Contents

Preface		**vii**
Chapter 1	Introduction	**1**
Chapter 2	Chitosan Glass Transition Temperature Controversy	**3**
Chapter 3	The Glass Transition and the α-Relaxation Process	**7**
Chapter 4	Dielectric Relaxation Processes in Polysaccharides	**9**
Chapter 5	Dynamic Mechanical Analysis (DMA) in Chitosan	**11**
Chapter 6	Materials and Experimental Methods	**13**
Chapter 7	Results and Discussion	**23**
Chapter 8	Conclusions	**39**
References		**41**
Index		**45**

PREFACE

In this chapter thermal relaxation properties of chitosan neutralized and non-neutralized films will be reported as a function of water content using dynamical mechanical analysis and dielectric spectroscopy in the temperature range from 20 to 250°C. Three relaxation processes have been observed in different temperature and frequencies ranges. For the first time, the low frequency α-relaxation associated with the glass-rubber transition, which relate to a plasticizing effect of water, has been detected by this technique in both chitosan forms in the temperature range 20-70°C and for moisture contents between 0.5 to 10 wt %. On this basis, the glass transition temperature was estimated in the range 86-102°C which shifts to higher temperature with decreasing moisture content and the glass transition vanishes in a dry material. A second low frequency relaxation was observed from 80°C to the onset of thermal degradation (240°C) and identified as the sigma-relaxation often associated with the hopping motion of ions in the disordered structure of the biomaterial. This relaxation exhibits a normal Arrhenius-type temperature dependence with activation energy of 86-88 kJ/mol and it is independent of water content. The non-neutralized chitosan possess higher ion mobility than the neutralized one as determined by the frequency location of the σ-relaxation. A high frequency (10^4 -10^8 Hz) secondary β-relaxation in neutralized and non-neutralized chitosan, related to side group motions by means of the glucosidic linkage is observed in the temperature range 20-120°C and moisture contents less than 3 wt % with Arrhenius activation energy of 46.0-48.5 kJ/mol. In the films with higher moisture contents this relaxation is not well resolved due to a superposition of two relaxation processes: beta and beta wet relaxations that merge into one common β-relaxation process.

Chapter 1

INTRODUCTION

The knowledge of thermal behavior of a polymer is of scientific and technological interest when temperature is a processing variable in device fabrication. Among several physical properties dependent on temperature, a glass transition can be used to characterize a property of a polymeric material. The glass transition temperature (T_g) constitutes the most important mechanical property for all polymers; since at this temperature the polymer goes from hard-glass like state to a rubber like state. This property is a specific feature for polymers amorphous domains and upon synthesis of a new polymer, glass transition temperature is among the first properties measured since the state parameters will exhibit an abrupt change within the range of the glass transition temperature. Hardness, volume, Young's modulus, percent elongation-to-break, viscosity, density, thermal expansion coefficient and heat capacity are some of the physical properties that undergo a drastic change at the glass transition temperature. So any instrument that can detect the thermal, mechanical or dielectric changes can theoretically be used to measure the T_g of a material. The glass transition itself is currently not well understood theoretically and the nature of the glass transition can be described as one of the most serious unsolved and challenging problems in condensed matter physics [1]

In the last four decades, it has been controversial as to whether chitin and its main derivative chitosan exhibit a glass transition temperature (T_g). In semi-crystalline polymers, such as chitin and chitosan, a glass transition temperature characteristic of the amorphous material can usually be detected. However, it is controversial as to whether polysaccharides exhibit a glass transition temperature (T_g) [2]. For chitosan, some authors using several techniques, that include differential scanning calorimetry (DSC) and dynamic mechanical thermal analysis (DMTA), have reported T_g values from 20 to 222°C [3-10], whereas others do not observe the glass transition [11-14]. DMA technique has been a common tool to analyze relaxational process in chitosan, however, these studies are performed at one low fixed frequency and the identification with T_g is tenuous, thus the nature of each relaxation process is also controversial and several different values have been assigned.

In polymers the glass transition phenomenon has been related to the dielectric α-relaxation processes through the Vogel-Fulcher-Tammann (VFT) equation [15]. Dielectric spectroscopy is a significant technique used to investigate the relaxation properties of materials. The main advantage of dielectric technique over DSC and DMA analysis that attempt to measure molecular dynamics is the extremely broad frequency range covered. Some authors have reported molecular dynamics analysis in chitosan by dielectric measurements with no evidence associated with a glass transition [11, 16, 17]. These studies reported the β-relaxation process related to local main chain motions at the high frequency side; and at higher temperature the so-called σ-relaxation produced by proton migration [16, 17]. It is well known that this biopolymer is highly hydrophilic and small amounts of water affect its molecular relaxations, this effect has been reported on wet chitosan since it exhibits an additional relaxation referred as the β_{wet}-relaxation [11, 16, 17]. Recently, molecular relaxations in chitin [18] and chitosan [2] have been investigated using dielectric spectroscopy, these results show evidence of the low frequency α-relaxation present in *dry annealed* films, this relaxation process is related to the glass transition and a glass transition temperature is assigned by fitting the experimental data to the Vogel-Fulcher-Tammann model.

On the other hand, the physical and chemical properties of this biopolymer can be significantly changed by the presence of small amounts of water [2, 16, 18, 19]; since chitosan has a strong affinity for water and therefore, may be readily hydrated forming macromolecules with rather disordered structures [20], a true understanding of hydration properties is essential for several applications in materials science, food industry and biotechnology [20]. Dielectric spectroscopy is a significant technique used to investigate the hydration properties of materials [2, 18, 19] and glass transition phenomena as well. That is why the objective of this book is to review the dielectric and DMA analysis to discern the nature of different molecular relaxation process and its relation ship to the α-relaxation process and the glass transition in hydrated and dry chitosan thin films.

Chapter 2

CHITOSAN GLASS TRANSITION TEMPERATURE CONTROVERSY

Chitosan possess good film-forming properties. Numerous publications have reported on studies of films made of chitosan [21]. Among others, an important parameter on films-forming processing is the glass transition temperature (T_g). It is important not only in optimizing manufacturing processes, but in understanding the reliability implications of exposure of the products to thermo-mechanical stresses as well. The application of the glass transition concept can be also a valuable aid in edible films research, as T_g affects their mechanical and barrier (gas, eater vapor) properties. The T_g is a key factor for deciding the usefulness of a polymer.

Regarding chitosan glass transition temperature, it have been main object of study during the last years, owing to the fact that there are a wide variety of assigned T_g values as it is shown in Table 1. Being a natural polymer, the source and method of extraction, cristallinity, molecular weight and deacetilation degree are some properties that can influence the T_g [20] and this could be the reason for the huge range of reported values.

The concept of T_g only applies to non-crystalline solids, which are mostly either glasses or rubbers. Non-crystalline materials are also known as amorphous materials. Amorphous materials are materials that do not have their atoms or molecules arranged on a lattice that repeats periodically in space. For amorphous solids, whether glasses, organic polymers, or even metals, T_g is the critical temperature that separates their glassy and rubbery behaviors. A small change in temperature T_g could result in pronounced changes in the mechanical, thermal and dielectric properties of amorphous materials.

Table 1. Summary of the Chitosan glass transition temperature (Tg) reported in literature

Author	Year	Technique	Tg (°C)	Observations
Chitosan Glass Transition Temperature assigned				
Kaymin et al.[22]	1980	DIL and DMA	55	Dried samples at 100°C and annealed at 196°C
Ogura et al.[23]	1980	DMA	140	
Pizzoli et al.[11]	1991	DMA and DS	130	Water induced ☐ d relaxation
Kim et al.[24]	1994	DMA	150	
Ratto et al.[3]	1995	DSC	34	Water content near to 0%
Arvanitoyannis et al. [10]	1997	DETA	110	
Guan et al. [25]	1998	DSC and DMA	90 and 105	
Dufresne et al.[26]	1999	DMA	No Tg	
Sakurai et al.[7]	2000	DSC and DMA	203	Measured after heating up to 180°C
Toffey et al.[27]	2001	DMA and DSC	60-93	DSC didn't show glass transition
Kittur et al. [12]	2002	DSC	No Tg	
Lazaridou et al. [21]	2002	DMA	95	DMA measurements demonstrate the plasticizing effect of water giving values near to 77°C for water content ca. 0.1 weight fraction
Zohuriaan et al. [28]	2004	DSC	No Tg	
Dong et al. [5]	2004	DIL, DMA, TSC and DSC	140-150	They found two more relaxations at 85 and 197°C. Tg values do not depend on deacetilation degree
Wu et al. [29]	2004	DMA	153	They report ☐-relaxation at 15-22°C
Mucha et al.[4]	2005	DMA	156-170	Tg values depends on deacetilation degree

Author	Year	Technique	Tg (°C)	Observations
Liu et al.[6]	2006	DSC	92	Reported value from first scan
Shieh et al. [30]	2006	DMA	150	One relaxation at 25°C attributed to the glass to rubber transition of water-plasticized chitosan
Quijada-Garrido et al.[31]	2007	DMA and DSC	85-99	Tg value depends on frequency. DSC showed no glass transition.
Rao et al.[8]	2008	DSC	222	Only one scan was performed
Mayachiew et al. [9]	2008	DSC	193 and 197	Tg value depends on drying technique.

Dilatometry (DIL).
Dynamic Mechanical Analysis (DMA).
Dielectric Spectroscopy (DS).
Dynamic Electrical Thermal Analysis (DETA).
Thermally Stimulated Current Spectroscopy (TSC).
Differential Scanning Calorimetry (DSC).

In Differential Scanning Calorimetry (DSC) analysis, the glass transition is defined as a change in the heat capacity as the polymer matrix goes from the glassy state to the rubbery state. This is a second order endothermic transition that requires heat to go through the transition, so in the DSC the transition appears as a step transition and not as a peak such as in a melting transition. DSC is the classic and "official" way to determine T_g even though in some cases there are polymeric materials that do not exhibit a sharp T_g by DSC, this had been the case of chitin and chitosan as well as cellulose [24, 29]. On the other hand, in Thermal Mechanical Analysis (TMA) this transition is associated with a change in the free molecular volume and defines the glass transition in terms of the change in the coefficient of thermal expansion (CTE) as the polymer goes from glass to rubber state. Each of these techniques measures a different result of the change from glass to rubber. The DSC is measuring a heat effect, whereas the TMA is measuring a physical effect i.e. the CTE. Both techniques assume that the effect happens over a narrow range of a few degrees in temperature. If the glass transition is very broad it may not be seen with either approach.

From the practical point of view, fundamental information on the process ability of polymers is usually obtained through thermal analysis; it provides information of the main polymers transitions related to melting and glass-to-rubber transition to the crystalline and amorphous phases, respectively. In addition to the well established calorimetric techniques, experimental methods capable of revealing the motional phenomena occurring in the solid state have attracted increasing attention. In amorphous polymers, α-relaxation, as determined by Dielectric Spectroscopy (DS) and Dynamic Mechanical Analysis (DMA) corresponds to the glass transition and reflects motions of fairly long chain segments in the amorphous domains of the polymer involving long range motions. Relaxations at lower temperatures (labeled β, γ, δ...) are generally due to short range motions related to local movements of the main chain, or rotations and vibrations of terminal groups or other side chains. DS and DMA are well-established techniques for the measurement of thermal transitions including the glass transition; they are especially available in detecting T_g of a sample that cannot be observed by normal calorimetric measurements. For example, T_g of polymers having crosslinked network structure [32]. These two techniques were used in the present work for chitosan thermal analysis.

Chapter 3

THE GLASS TRANSITION AND THE α-RELAXATION PROCESS

When studying a polymer like chitosan on a large frequency/time scale, its response under a dielectric or dynamic stimulus could exhibits several relaxations. More over, the peaks could be usually broad and sometimes a superposed processes [2, 18, 33]. Molecular processes cover a broad frequency range and they are associated with the length scale of the conformational mobility in the polymeric chain. Relaxation processes in polymeric materials involves several relaxation process that go from secondary process that are very local motions (β, δ, γ, etc) to primary relaxation process due to segmental mobility involving co-operatives (α-process), or even relaxation processes involving large or complete polymeric segments [34]. The relaxation rate, shape of the loss peak and relaxation strength depend on the motion associated to a given relaxation process. In general, the same relaxation/retardation processes are responsible for the mechanical and dielectric dispersion observed in polar materials [34].

The α-relaxation is related to the glass transition and in general they are not well understood, and the real microscopic description of the relaxation remains being a current problem of polymer science [35]. However it is well accepted that the dynamics of the glass transition is associated with the segmental motion of chains being cooperative in nature [35], which means that a specific segment moves together with its environment. For most amorphous polymers in the α-process the viscosity and consequently the relaxation time increase drastically as the temperature decreases.

Therefore molecular dynamics is characterized by a distribution of relaxation times and by a non-Debye response. A strong temperature dependence presenting departure from linearity or non-Arrhenius thermal activation is a well description of the glass transition process, due to the abrupt increase in relaxation time as temperature decreases developing a curvature near T_g. This dependence can be well described by the Vogel-Fulcher-Tammann (VFT) equation [34, 35]:

$$\log f_{p\alpha} = \log f_{\alpha,\infty} - \frac{A}{T-T_o} \qquad (1)$$

where $\log f_{\alpha,\infty} = 10^{10}\text{-}10^{13}$Hz and A are constants, and T_0 is the so called ideal glass transition or Vogel temperature, which is generally 30-70°K below T_g [15, 35].

In general the α-process is well defined in the frequency domain and shows a relatively broad and asymmetric peak. Several functions like the Cole-Cole and Cole-Davison are able to describe broad symmetric and asymmetric peaks in the frequency domain and the most general one is the model function of Havriliak and Negami (HN function) [35 pp]. In order to investigate the nature of chitosan relaxation processes, the fitting of the complex permittivity in *dry annealed* and *dry* neutralized and non-neutralized samples were carried out using the well-known HN empirical model [15, 22]: $\varepsilon^* - \varepsilon_\infty = \frac{(\varepsilon_s - \varepsilon_\infty)}{\left[1+(j\omega\tau)^\alpha\right]^\beta}$, where ε_s-ε_∞ and ε_∞ are the dielectric relaxation strength and the dielectric constant at the high frequency limit, respectively. The exponents α and β introduce a symmetric and asymmetric broadening of the relaxation. This general equation includes especial cases of relaxations processes for Debye (α and β=1), Cole-Cole (α=1) and Davison-Cole (β=1).

Chapter 4

DIELECTRIC RELAXATION PROCESSES IN POLYSACCHARIDES

To investigate molecular motions by Dielectric Spectroscopy (DS) both, the repeating unit of a polymer and the attached side groups must own a permanent dipolar moment to look into dipolar fluctuations in broad temperature and frequency ranges, this requirement is well accomplished by chitosan. Dielectric spectroscopy of polysaccharides has been considered controversial by many scientists, up to now [16]. Different molecular groups of a repeated unit of a polymer are separated by dielectric relaxation spectroscopy with respect to the rate of its orientation dynamics [16].

Einfeldt et al.[16] developed an exhaustive study of dielectric spectra of a great variety of polysaccharides including dry and wet cellulose-based materials, different wet starches and also derivatives of cellulose and starch. They observed different modes of relaxation processes in the sub-T_g range assigned as γ, δ, β_{wet}, β and σ relaxation processes involving side group motion, water effect and hoping motion of ions (these relaxation processes are resumed and extensively described elsewhere [16]). However they do not draw any discussion about the primary α-relaxation associated to the glass transition temperature since no evidence was found for a dynamics with VFT temperature dependence; the typical feature for the glass transition dynamics.

DS studies on chitosan are scarce and ambiguous. Even though the scientific and technologic importance of this polymer only a few papers exist concerning to its molecular dynamics. Pizzoli et al.[11] showed low- and high-temperature relaxations in dry chitosan (they called "dry chitosan" to vacuum annealed up to 180°C samples). Below room temperature, they observed a frequency dependent secondary relaxation ascribed to local motions of small molecular units and assigned as γ relaxation. Whereas at high temperature (above 140°C) they suggested the occurrence of a second relaxation, concluding that their results bring no evidence in favor of the glass transition attribution. On the other hand, Viciosa et al.[17] reported two main Arrhenius type relaxations processes in neutralized and non-neutralized chitosan; process I found below 0°C owing the characteristics of a secondary relaxation process related with local chain dynamics, and process II at higher temperatures correlated with dc conductivity.

Viciosa et al.[33] also reported the β-wet process located at temperatures below 0°C in wet samples, vanishing after heating to 150°C. Finally, Einfeldt et al.[16] found a β-relaxation in β-chitin and chitosan in the low temperature range with activation energies of 44.7 and 47.8 kJ/mol respectively. They also performed high temperature studies on chitin and chitosan [36], however, they do provided activation energies values for the σ-relaxation in these biopolymers.

Chapter 5

DYNAMIC MECHANICAL ANALYSIS (DMA) IN CHITOSAN

Dynamic mechanical analysis (DMA) is a thermal analysis technique that measures the properties of materials while they are deformed under periodic stress [37]. Since polymers are viscoelastic materials, i.e. they simultaneously exhibit solid-like and liquid-like properties; they are by definition time-dependent. They exhibit the properties of a glass (high modulus) at low temperatures and those of a rubber (low modulus) at higher temperatures. The scanning of temperature during a DMA experiments provides information about the glass transition or α-relaxation, so the T_g can be measured by DMA. DMA can also be used to investigate the frequency dependent nature of the transition; since T_g has a strong dependence on frequency DMA can also resolve secondary transitions, like β, γ, and δ transitions, in many materials that the DSC technique is not sensitive enough to pick up.

As mentioned above, in chitosan a wide variety of T_g values have been reported over the past years, basic studies were conduced on the molecular motion and thermal relaxation behavior of chitosan using the DMA technique, some of them taking into account moisture content influence on their physical properties. Ogura et al.[23] showed the dynamical mechanical behavior of dry chitosan, two loss peak assigned to the γ-relaxation and α-relaxation where observed. They proposed a T_g for dry chitosan around 140°C. Guan et al. [25] found two peaks in the DMA spectra of neutralized chitosan films; at ca. -55°C and 105°C, ascribe to the β and α-relaxations respectively. Ratto et al. [3] showed a low amplitude transition centered at 90°C assigned to local chain motion. Meanwhile, Toffey et al. [27] assigned a T_g= 60-93°C depending of the acid used to form the film, they also showed an additional low temperature relaxation at -10°C designed it as the β-relaxation.

Lazaridou et al. [21] observed the moisture content plasticizing effect on the glass transition temperature of chitosan and estimated a T_g ca. 95°C for dry chitosan. While Yu-Bey Wu et al. [29] observed two tan δ peaks; one attributed to the β-relaxation at 15-22°C and another at 153°C designed as the α-relaxation. As well as Quijada-Garrido et al. [31] that reported a β-relaxation at -30 to -12°C and an α-peak at 85-99°C depending on frequency. Ahn et al. [38] assigned 161°C as the α-relaxation peak and a peak around 60°C to what they describe as a water-induced β-relaxation. Whereas, Sakurai et al. [7] estimate the T_g to be 203°C by the tan δ curve after a first heating it up to 180°C, even though the onset of chitosan thermal degradation is around 170°C [39]. Neto et al. [20] observed a main event at 50°C, in the abstract they ascribe this relaxation to the glass transition; however in the manuscript they do not dare to say that it is the glass transition temperature.

Dong et al.[5] by four different techniques including DMA, assigned a T_g value of 140-150°C, they observed three more peaks at ca. 85°C, 197°C and 200°C attributed to a water-induced relaxation, an unknown α transition and decomposition respectively. M. Mucha et al.[4] recognized a broad tan δ peak at 156-170°C as the α-relaxation at T_g, they also found three more events assigned as T_1=-21°C, T_2=24°C and T_3=43°C that move to higher temperatures at T_1=-12°C, T_2=32°C and T_3=45°C in preheated samples. T_1 was assigned as β-relaxation, associated with local motions of side groups and T_2 and T_3 to strongest complex relaxations occurring at ambient temperature ascribed to structural reorganization of packing of chitosan molecules due to an increase of residual water mobility.

Shieh et al. [30] found two peaks, one at 25°C and another near to 150°C, the occurrence of the low-temperature relaxation was attributed to the glass-to rubber transition of the water plasticized chitosan, and the relaxation peak near 150°C to the glass transition of large-scale cooperative molecular motions in the amorphous phase.

Even though the wide variety of T_g values reported, there are some authors that found no glass transition evidence in chitosan by DMA measurements [11-14, 39-41].

Chapter 6

MATERIALS AND EXPERIMENTAL METHODS

NEUTRALIZED AND NON-NEUTRALIZED CHITOSAN FILMS PREPARATION

Chitosan medium molecular weight (Mw= 150,000 g/gmol) of 76 and 82 % of degree of deacetylation (DD) reported by the supplier and calculated according to [42], was purchased from Sigma-Aldrich. Acetic acid from J.T. Baker was used as received without further purification. Chitosan films were obtained by dissolving 1 wt % of chitosan in a 1 wt % aqueous acetic acid solution with subsequent stirring to promote dissolution. Chitosan films were prepared by the solvent cast method by pouring the solution into a plastic Petri dish and allowing the solvent to evaporate at 60°C. Chitosan films prepared from acetic acid solution have the amino side group protonated (NH_3^+ groups); therefore the films need to be neutralized. The films were immerse into a 0.1 M NaOH solution during 30 min and washed with distilled water until neutral pH, a subsequent dried step in furnace at 130°C for 14 hr was needed. The results of both *neutralized* as well as *non-neutralized* chitosan films are shown.

ELECTRODE PREPARATION FOR DIELECTRIC MEASUREMENTS

A thin layers of gold was vacuum-deposited onto both film sides to serve as electrodes have been prepared by using a sputtering system (Plasma

Sciences Inc.) with a gold target (purity 99.99%) and Argon as gas carrier. With a gas pressure set to 30 mTorr and voltage set to 0.2 kV. Sputtering time was 4 minutes onto each side. Rectangular small pieces (5 mm × 4 mm) of these films were prepared for dielectric measurements. The contact area and thickness were measured with a digital calibrator (Mitutoyo) and a micrometer (Mitutoyo), respectively. This method allows obtaining films of ca. 10 μm of thickness.

INFRARED MEASUREMENTS

Chemical analysis and degree of deacetylation calculation of chitosan (CTS) and CTS acetate/SN composites was performed by Fourier-Transform Infrared spectroscopy (FTIR) on a Perkin Elmer spectrophotometer model Spectrum GX, using an ATR accessory in the range 4000-650 cm-1, resolution was set to 4 cm-1 and the spectra shown are an average of 32 scans. Chitosan samples prepared in the forms of potassium bromide (KBr) disk and film were studied.

THERMAL MEASUREMENTS

Free water content was determined by thermogravimetric analysis (TGA). TGA curves were obtained using a Mettler Toledo apparatus, model TGA/SDTA 851e, using a sample mass of ca. 3 mg and an aluminum sample holder under argon atmosphere with a flow rate of 75 mL/min. Heating rate was set to 10°C/min. For dielectric measurements, three types of films have been studied: (1) *Wet* chitosan, samples prepared under normal ambient conditions (25% humidity, at 20°C); (2) *Dry* chitosan, samples heated at 120°C for 24 h then cooled at ambient conditions for 10-15 min prior to measurement; and (3) *Dry annealed* chitosan, *dry* samples with an additional annealing at 120°C for 1 hour in the experimental impedance cell followed by cooling to 30°C under vacuum. After this conditioning, dielectric measurements are taken up to 250°C. The annealing time of 1 hour was chosen because further annealing did not change resistance of samples.

Regarding DMA measurements, *wet* samples corresponds to samples under normal ambient conditions, *dry* samples were heated at 75°C before weight loss and DMA measurements and *dry annealed* samples were heated at 150°C before weight loss percent and DMA measurements.

Modulated differential scanning calorimetry (MDSC) measurements were performed in a Q100 TA Instruments calorimeter, using a sample mass of ca. 3-4 mg. Heating rate was set to 3°C/min with a modulation amplitude and frequency of ± 1°C and 1/60 s^{-1} respectively. Standard aluminum pans were used within a nitrogen atmosphere flowing at a rate of 80μL/min. An empty pan was used as reference. First and second scans were analyzed.

DIELECTRIC MEASUREMENTS

Dielectric measurements in the frequency range from 10^{-1} to 10^6 Hz were carried out using Solartron 1260 impedance gain-phase analyzer with 1294 Impedance interface and in the frequency range 40 Hz-100 MHz using an Agilent Precision Impedance Analyzer 4294A. The amplitude of the measuring signal was 100 mV. The Agilent 4294A was calibrated for fixture compensation according to [43] when connecting a direct with home-made vacuum cell to the impedance analyzer port. It includes: open, short and load compensation with the characteristic impedance of 100 Ω used. Three samples: *wet*, *dry* and *dry annealed* previously defined were studied. To ensure entire water removal for the *dry annealed* samples, heating for 1 h at 120°C is done prior to dielectric measurements.

An in-house impedance two contacts vacuum cell was used in conjunction with a Watlow's Series 982 microprocessor with ramping temperature controller for all dielectric measurements from 20°C to 250°C, which was programmed to produce a constant heating rate of 3°C/min between certain measurements temperature. Each sample was kept for 3 min at each temperature to ensure thermal equilibrium.

DYNAMIC MECHANICAL ANALYSIS (DMA)

Dynamic mechanical analysis was made with a Ta Instruments Dynamical Mechanical Analyzer model RSA III. The heating rate was 5°C/min at 0.1 HZ frequency and an initial strain 0.1%.

EXPERIMENTAL DATA PRE-TREATMENT

Figure 1 shows the complex Cole-Cole plot for *dry annealed* chitosan films. It exhibit characteristic semicircles at high frequencies and a quasi-linear response at low frequencies, the linear response at low frequencies can be associated with interfacial polarization in the bulk films and/or surface and metal contact effects [18, 44]. To analyze the dielectric relaxation of chitosan films is necessary to understand the nature of the low frequency part of dielectric spectrum.

To test the influence of gold contact on the chitosan dielectric spectra, a non-symmetrical contact array was prepared as reported elsewhere [2]. Window insert of Figure 1 shows dielectric measurements on wet Au-chitosan-Au sample with different applied bias voltage at 23°C. The bias increase leads to a reduction of barrier resistance that changes the low frequency contribution of dielectric spectra (which occur at large values of Z') as shown in window insert of Figure 1. As can be seen, a deviation from a semi-circle is seen at both temperatures and the response depends on the bias voltage. This behavior is a good indication of contact polarization effects related to the (partial) blocking of charge carriers at the film/electrode interface [45]. This low frequency part of the electrical response is easily influenced by imperfect contact between the metal electrode and the sample.

In addition to this electrode polarization, interfacial polarization effects are observed in the high temperature range (>120°C) for all chitosan films. This effect manifests as a "bulge" on the semicircle. Figure 1 illustrates this effect at 125, 135 and 145 °C. According to the classical model the appearance of the interfacial polarization in dielectric spectrum can be observed as appearance of additional semicircle [2, 18]. This is a common form of a discontinuity occurring in an inhomogeneous solid dielectric associated with internal interfaces; it is well-known as the Maxwell-Wagner-Sillars (MWS) relaxation. In polysaccharides and biopolymers [16, 17] the interfacial polarization (MWS polarization) was observed in low frequency and high temperature ranges. In the low frequency range both contact and interfacial polarizations were observed in all samples. These polarizations have to be carefully considered since it is important to take into account only a so-called depressed semicircle that does not include contact and interfacial polarization effects. For most polysaccharides a commonly used plot of admittance versus frequency does not reveal the appearance of the extra semicircle related to interfacial polarization [18]. It is noteworthy that data treatment proposed in this study allows one to identify and separate these two processes (contact and interfacial polarization effects) and consequently before fitting dielectric data to models, this low frequency data needs to be discarded because a model-based analysis can be misleading if appropriate contact and interfacial polarizations are not considered.

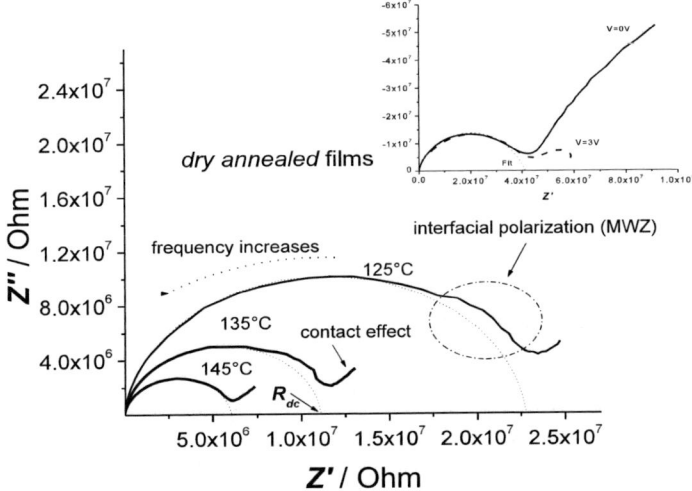

Figure 1. Complex dielectric spectrums of dry annealed chitosan films. Window insert: Dielectric spectra at different bias applied voltage. Contact effect is observed when applied voltage is changed from 0V to 3V in the quasi linear response (low frequency side), in a film tested at 25°C.

MOISTURE CONTENT EFFECTS ON CHITOSAN MOLECULAR RELAXATIONS AND *DC* CONDUCTIVITY CORRECTION

Chitosan is a hydrophilic polysaccharide and it is well known that moisture content has a significant influence on its physical properties [16, 18]. A true understanding of hydration properties is essential for several practical applications in materials science, food industry, biotechnology, etc [46]. Chitosan moisture content is affected by the number of ionic groups in the material as well as their nature. Hydroxyl and amine groups present in chitosan are important binding sites for water molecules. The glass transition phenomenon could be affected by moisture content, since it can work as a plasticizer [2]. Plasticization occurs in the amorphous region only, such that the degree of hydration is quoted as moisture content in the amorphous region.

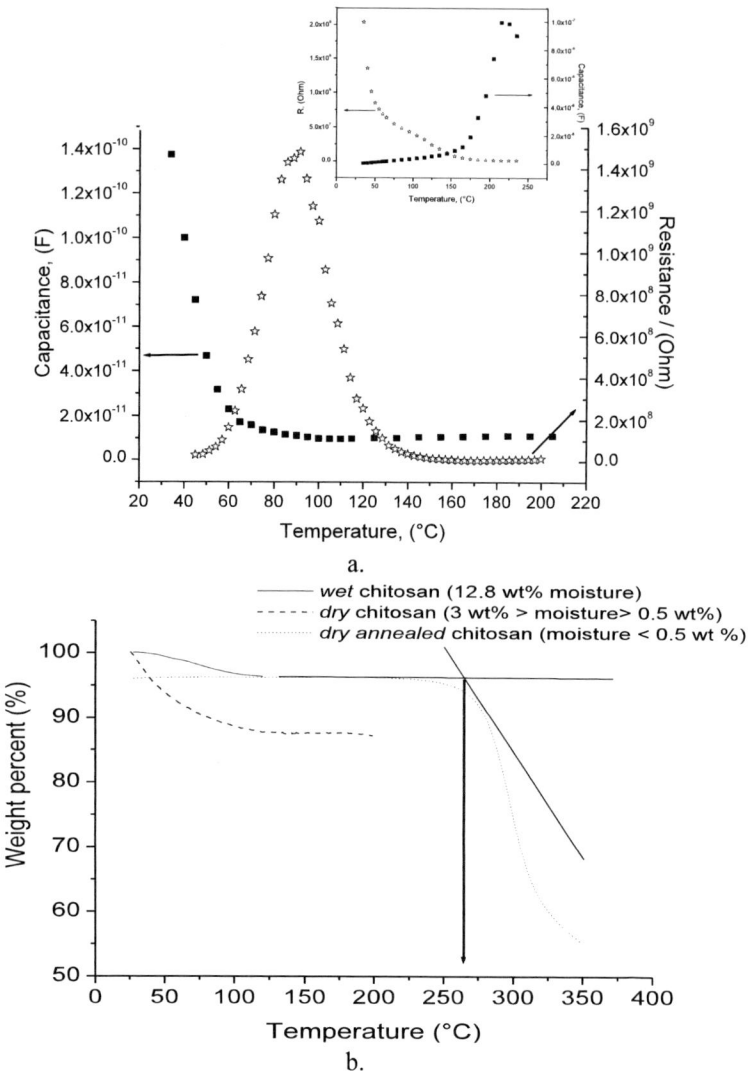

Figure 2. a) Resistance (open stars) and capacitance (solid squares) as a function of temperature for wet film. Thermogravimetric response is also shown. Note, that as temperature increases resistance increases due to water evaporation and finally decreases. Window inset shows the opposite behavior for *dry* films after water evaporation by annealing. b) Thermogravimetric analysis of solvent-cast chitosan films for w*et, dry* and *dry annealed* films.

Figure 2a shows the dependence of *dc resistance* and *capacitance versus temperature* for *wet* chitosan (12.8 wt% moisture content). It can be observed that in the temperature range 20-70°C the resistance increases and capacitance decrease. This behavior is ascribed to water modification effect on the relaxation mechanism of the matrix, since water has a lower resistance and higher dielectric constant. When water is present, biopolymer's resistance decreases; if temperature increases water evaporates and resistance increases (weight loss about 12.8 wt% in the TGA measurements is registered, see Figure 2b). When water content is below 3 wt% (*dry* and *dry annealed* samples of Figure 2b), the real dielectric behavior of chitosan is revealed. Films with near zero moisture content exhibit higher resistances at room temperature and lower resistance as temperature increases as it is shown in window insert of Figure 2a (*dry annealed* film). In summary, to obtain the dielectric behavior of pure chitosan without moisture influence, it is necessary to evaporate it by heat treatment and an additional annealing at 120°C, otherwise its evaporation mask electrical properties of the biopolymer.

For *dry* films, we emulated dielectric measuring conditions for TGA measurements: after overnight annealing in an oven at 120°C, these films were handled in ambient conditions for 10-15 min for TGA measurements (time to weight samples for TGA ≈ time for sample handling prior to dielectric measurements). The weight loss, and therefore moisture content is ca. 3.0%, indicates that chitosan reabsorbs water readily during the 10-15 min handling from the oven to the vacuum cell. So a second heat treatment in the vacuum cell prior to dielectric measurements is needed to obtain nearly *zero* weight percent of moisture. These films were reheated at 120°C in the impedance vacuum cell for 1 hour (*dry annealed* films moisture content ≈ 0.5 wt %).

Besides the variations of dielectric properties of a hydrated material due to the polar water molecules themselves, there is a second one due to the modification of the various polarization and relaxation mechanisms of the matrix material itself by water [19]. Furthermore, in the low-frequency region of measurements there is a third contribution often ignored in work dealing with high-frequency measurements, arising from the influence of moisture on conductivity and conductivity effects. The increase of electrical conductivity of the sample is the major effect present in wet samples, dielectric response is often masked by conductivity, and it superposes the dielectric processes in the loss spectra, this effect demands a conductivity correction of the dielectric loss spectra [15] because it strongly affects the modified loss factor ε". In this case it can be expressed as:

$$\varepsilon^{"} = \varepsilon^{"}_{exp} - \frac{\sigma_{dc}}{\omega\varepsilon_0} \qquad (2)$$

where $\varepsilon^{"}_{exp}$ is the experimental loss factor value, σ_{dc} ($\sigma_{dc} = d/(R_{dc} \times S)$) is direct current conductivity, d and S are thickness and area of sample respectively, $\omega=2\pi f$ (f is frequency), and ε_0 is the permittivity of vacuum. As a general rule for polymers, σ_{dc} is determined from fitting of real component of the complex conductivity ($\sigma_{dc}=\sigma_0 f^{\,n}$, where σ_0 and n are fitting parameters) measured in the low frequency range where a plateau is expected to appear [36]. However, in our samples this plateau is not resolved, as a consequence polarization and contact effects cannot be discerned and the correct *dc* conductivity cannot be calculated by this method, so we use an alternative procedure to circumvent this problem. In the Z'' versus Z' plot, the values of *dc* resistance R_{dc}, and the corresponding conductivity $\sigma_{dc}=d/(R_{dc}\times S)$, have been obtained from the extrapolation of the high frequency semicircle to the Z' axis as shown in Figures 1. On the other hand, Figure 3 shows experimental data before and after *dc* conductivity correction for *dry annealed* neutralized chitosan, it can be seen the high effectively of the method previously described to discard conductivity effects; since it reduces significantly the conductivity contribution in the low frequency side of the spectra and at the same time allows disclosing the low frequency relaxations.

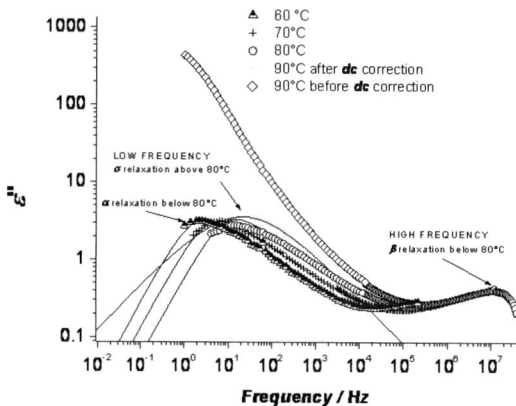

Figure 3. Dielectric spectra for neutralized chitosan. Note: low frequency relaxations are disclosed after dc conductivity correction. Low and high frequency relaxations can be fitted independently.

The primary α-relaxation associated with the glass transition, should appear below the kHz region [19]. After *dc* correction two different relaxations can be identified as seen in Figure 3 in the low frequency range (10^0-10^3 Hz) and one more above 10^5 Hz (high frequency side). This high frequency relaxation was previously reported as the secondary β-relaxation observed in the 10^4-10^8 Hz frequency range. The two relaxation processes are well defined and separated in the frequency range, and they can be investigated and correlated as the low and high-frequency relaxation processes independently. In the low frequency side and above 70°C a low frequency relaxation with Arrhenius temperature dependence (as will be shown later) is revealed, and below 70°C a different relaxation with a non-Arrhenius temperature dependence that we designate as the primary or α-relaxation is disclosed. The same behavior was observed in non-neutralized films (nor shown here).

Chapter 7

RESULTS AND DISCUSSION

DIELECTRIC SPECTROSCOPY ANALYSIS

Low Frequency Relaxations

Once contact and interfacial polarization were discarded and *dc* correction was performed as described in previous section; high and low frequency relaxations were identified. The R_{dc}, and the corresponding *dc* conductivity σ_{dc} ($\sigma_{dc}= d/(R_{dc} \times S)$, where d is the film thickness and S the contact area) have been obtained from the intersection of the high frequency semicircle and the real-part axis on the impedance plane as it was shown in previous section and in Figure 1. The *log_{10} dc conductivity versus 1000/T* dependence is shown in Figure 4. A similar behavior (moisture content-wise) in neutralized and non-neutralized films is shown, the non-neutralized chitosan films (open circles Figure 4) own higher conductivity values due to higher mobility and/or number of conducting species (NH_3^+ groups) [17].

The analysis of the conductivity plot shows two clear relaxations separated by temperature, so they can be analyzed in two temperature ranges. The "*low-temperature relaxation*" from 20 to 70°C and the "*high-temperature relaxation*" disclosed from 70°C to the onset of degradation ≈ 210°C.

Figure 4 shows conductivity results for the two chitosan used in this case: with 76 DD% and 82 DD% and obtained from shrimp and crab shell respectively. The same behavior described above is disclosed in both samples; below 70°C the characteristic α-relaxation behavior is present, and the high-temperature Arrhenius type relaxation is revealed above 70°C until the beginning of thermal degradation about 210°C. The Vogel temperature calculated from the fitting of our experimental data to the VFT relationship seems to be independent of the DD% and the raw material used for chitosan synthesis, since the same Vogel temperature ($T_0 \approx 269$ K) from the VFT fit to the α-relaxation (explained later) is obtained in both chitosan samples (see Figure 4).

Regarding the influence of DD% on the glass transition temperature, M. Mucha et al.[4] studied four different samples with 59, 67, 78 and 86 DD%, they stated that the temperature which they designed as the T_g of chitosan decreases from 167 to 156 °C with increasing DD% of chitosan. While Dong et al.[5] found no influence of DD% on T_g for chitosan with 46, 64, 71, 91 and 100 DD %. They stated that the glass transition should not be influence by DD because the α-relaxation belongs to the motion of segments in the main chain and if this would happen, this relaxation process should be present due to side group (i.e., acetamino or amino group) motion.

Chitosan with 82 DD% exhibits lower conductivity values compared to that with 76 DD% (see Figure 4), this difference is ascribe to the neutralization process, since the sample corresponding to 76 DD% is in the non-neutralized form (experimental section for details). According to [17], this higher conductivity values indicates higher mobility due to the presence of more conductive species in the form of NH_3^+ groups. As can be seen, both chitosan forms show the same relaxation behavior in the whole temperature-range. Therefore the influence on thermal relaxation of the degree of deacetylation and the raw material used for its synthesis can be rule out for further analysis. The following molecular relaxations analysis presented here includes neutralized and non-neutralized chitosan with 82 DD%.

Low Temperature and Low Frequency Relaxation

Let us focus on the 20-70°C temperature range, this relaxation is strongly affected by moisture content; a decrease in conductivity as temperature increases is clearly shown in *wet* samples (moisture content > 3 wt% calculated by TGA analysis), this effect is related to water evaporation and vanishes above 100°C disguising the real dielectric properties of chitosan above this temperature. For moisture contents below 3 wt% and above 0.05 wt% (*dry* films), a non-Arrhenius behavior emerges while for moisture content below 5 wt% (*dry annealed* samples) this low temperature relaxation vanishes after the second heat treatment (the discussion of this issue is differed for further explanation).

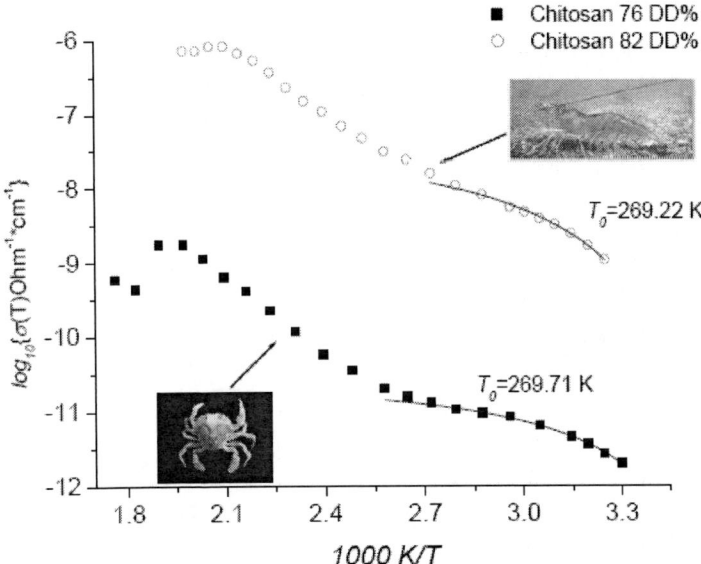

Figure 4. Temperature dependence for chitosan dc conductivity of neutralized (solid squares) and non neutralized (open squares). Note: the same Vogel temperature is obtained for both chitosan with different DD%.

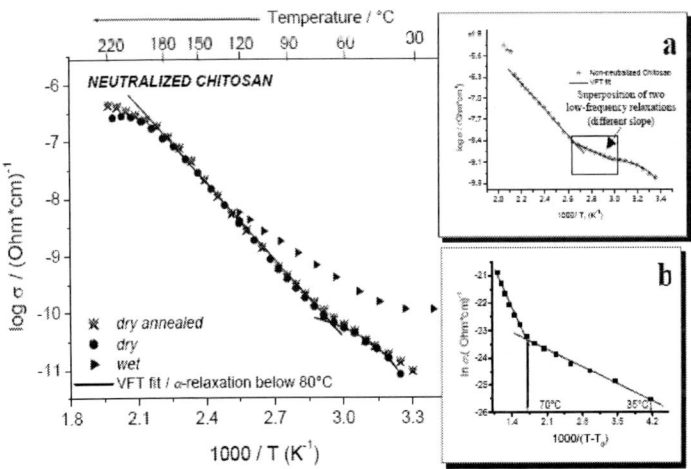

Figure 5. $\log_{10}\sigma$ versus $1000/T$ for *wet*, *dry* and *dry annealed* neutralized chitosan. Thermal relaxations are strongly affected by free-moisture content, α-relaxation is observed in the 30-70°C temperature range. Window insert: Non-neutralized chitosan, a superposition of two low-frequency relaxations is observed.

This non-linear dependence disclosed in the 20-70°C is the typical trend of the α-relaxation behavior related to the dynamic glass transition. The temperature dependence of conductivity and relaxation time calculated from HN fitting (Figures 5 and 6) are well described by the VFT equation $\sigma = \sigma_0 \exp[-DT_0/(T-T_0)]$ and $\tau = \tau_0 \exp[DT_0/(T-T_0)]$ [15]. There is an excellent agreement between the Vogel temperature calculated by conductivity and relaxation time independently (T_0 =284°K see Table 2). Note, that non-neutralized films exhibit higher conductivity than the neutralized ones, which is consistent with the excess number of protons in the non-neutralized films (NH_3^+ groups are present). Also, the T_0 value is independent of the presence of NH_3^+ groups.

The plot of log_{10} conductivity versus $1/(T-T_0)$ describes a straight line of negative slope in the 35-70°C temperature range. This linearity is a good indication of the α-relaxation; therefore all experimental data that fulfill this condition (Window insert of Figure 5b) belong to the α-relaxation process range (35 to 70°C).

The activation plot of relaxation time for the non-Arrhenius dependence present in the low frequency range below 70°C (Figure 6) as well as for secondary relaxations with Arrhenius behavior was obtained using the empirical Havriliak and Negami (HN) relation. For α-chitin (chitosan precursor) as well as for various cellulose-based materials, starches and non-polymeric glass forming liquids, it was found that the activation energies from the dc-conductivity plot and the dielectric α and σ-relaxation times (or α-peak frequency) are well correlated [36, 47, 48]. In glass-forming liquids, at high temperatures, both dc-conductivity and relaxation time show an Arrhenius behavior with the same activation energy. Below a cross over temperature (T_c) a Vogel-Fulcher-Tammann (VFT) behavior was observed [48].

Table 2. Parameter values for the VFT model in the range 30-80°C. The VFT model is generally used to describe α-relaxation. Both dc conductivity and relaxation time calculations are shown

α-relaxation VFTH Parameters	τ_0 or σ_0	D	T_0, K
NEUTRALIZED CHITOSAN			
Relaxation time	3.3×10^{-2}	0.32	284.3
dc Conductivity	3.1×10^{-10}	0.30	284.0
NON-NEUTRALIZED CHITOSAN			
Relaxation time	4.2×10^{-3}	0.43	283.9
dc conductivity	1.7×10^{-9}	0.10	283.9

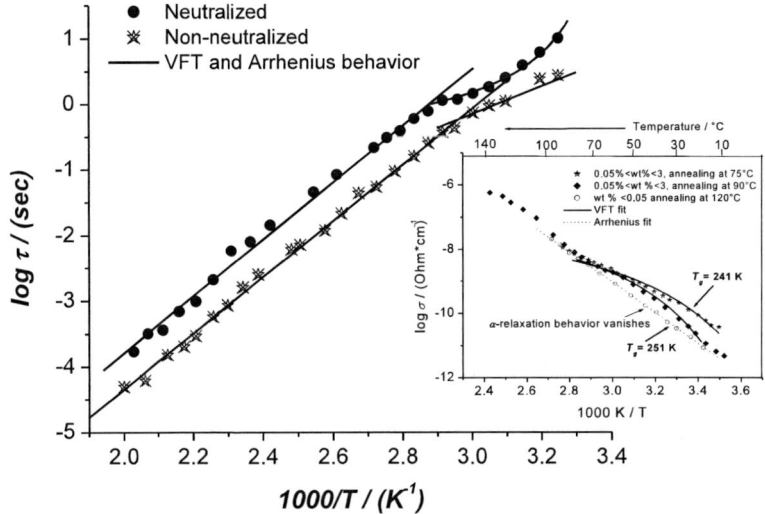

Figure 6. log_{10} *relaxation time versus 1000/T*. Low and high temperature relaxations. Lower relaxation times due to higher mobility conferred by NH^+_3 groups are demoted in non-neutralized films. Window insert: the α-relaxation is plasticized by water; T_0 shifts to higher values as moisture content decreases.

The two different temperature dependencies described above for glass-forming liquids are present in chitosan, and the similarity between *dc* conductivity and relaxation time for the two low frequency relaxations is clearly observed in Figures 5 and 6. Both dependences (*dc* conductivity (σ_{dc}) and relaxation time (τ) versus *1/T* plots) show the same features: an Arrhenius type relaxation will yield a straight line above 70°C (the crossover temperature (T_c) that separates the non-linear process from the linear process) whereas a non-Arrhenius relaxation will manifest as a curved line that suggests a VFT type or glass transition below 70°C in *dry* samples. For *wet* samples the decrease of conductivity as the temperature is increased from 20 to 70°C is likely due to the motion of water-polymer complex since water could be modifying the relaxation mechanism of the matrix material. And finally the vanishing of the non-Arrhenius behavior for moisture contents lower than 0.05 wt% labeled as *dry annealed* samples.

The Vogel temperature is the apparent activation temperature of the α-relaxation [15]. As proposed by a number of authors for different polymer systems using dielectric spectroscopy [15, 18, 49, 50] in the present study, we take $T_g = T_0 + 50$ K, since many polymers T_0 is usually 30-70K lower than T_g [20, 44]. We estimated a T_g value for neutralized and non-neutralized chitosan of 61°C (for moisture and heat treatment conditions explained before). The strong effect of moisture content on the α-relaxation process is also evidenced in window insert of Figure 6, since for *dry* samples the glass transition temperature shifts to higher values as moisture content decreases; it is possible to identify this effect when different annealing treatments at 75, 90 and 120°C are performed during measurements on the same sample. This shifting of the glass transition temperature to higher values as moisture contents decrease, points out a plasticizing effect of water on chitosan glass transition. Unfortunately, the accurate moisture content in the 0.05 to 3 wt% range according to the annealing treatment at different temperatures is difficult to determine since all the measurements are performed in a vacuum cell. However, it is noteworthy that moisture contents between 0.05 and 3 wt% are needed to distinguish the α-relaxation process by dielectric measurements before it vanishes at 0.05 wt% moisture content with the 120°C annealing treatment (see Figures 5, 6 and window insert of Figure 6).

Chitosan has functional groups like hydroxyls, amines and amides, which can act as hydrogen bond acceptors or donors. For this reason chitosan can be bonded or linked with hydrogen bond donors or acceptor compounds like water [31]. In the case of the glass transition temperature, plasticization occurs in the amorphous region only, such that the degree of hydration is quoted as moisture content in the amorphous region [51]. According to [52], the water sorption mechanism is composed of two main steps: water sorption on polymer sites and water clustering surrounding the first absorbed water molecules. Our results show that the α-relaxation process is strongly affected by the moisture content in the films as it is shown in Figures 5 and 6; while at percents higher than 3 wt% this process cannot be distinguished because of the free water effect, at lower percents (<3 wt% freezable bond water) a glass transition temperature can be assigned by the motion of a water-polymer complex in amorphous regions, a glass transition temperature can be assigned depending upon moisture content limited to be between 0.05 and 3 wt%. According to [31], the glass transition must be interpreted as torsional oscillations between two glucosamine rings across glucosidic oxygen and a cooperative hydrogen bonds reordering. Water sorption in hydrophilic polymers is usually a nonideal process leading to plasticization [52]. Water leads to an increase in the amounts of hydrogen bonds producing an increase in cooperative motion. Water hydrogen bonds between chitosan chains and increases *free volume*. When this happens chains can slide past each other more easily, so the time scale of the cooperative motion matches that of the experiment and the glass transition can be detected.

Nevertheless, if sample moisture is minimized (<0.05 wt %), chains are able to interact with each other giving rise to a denser packing. Thereby the mobility of polymer chains decrease and the glass transition phenomenon is not easily detected. Because of the absence of water, the glass transition temperature could be shifting to temperatures above 70°C (more rigid backbone). A similar effect than Figures 5 and 6 (20-70°C temperature range) at different moisture contents was observed in the hydrogel poly(hydroxyethyl acrylate) (PHEA) [19]. The results reported in this work [19] suggest that at h ≤ 0.21≈ 17.3% water (where h is defined as grams of water per gram of dry sample) conductivity is governed by the motion of the polymeric chains, whereas at h ≥ 0.29 (≈22.5 % of water), conductivity occurs through a separate water phase. At the highest moisture contents of 0.29 and 0.46 (≈ 31.5 % of water) the *dc* conductivity dependence changes from VFT type to Arrhenius type [19].

When water is minimized (0.05 wt% of water, Figure 6) and plasticization is absent, the glass transition temperature could be shifting to higher temperatures. At this point (above 70°C), another molecular relaxation takes place and the α-relaxation is weaker and cannot be observed because of the *high temperature* relaxation effect, this is clearly observed in *dry annealed* chitosan below 70°C (Figures 5 and 6) by a change of the VFT behavior to a linear one with a different slope than that for the *high temperature* process. This "new" slope is a behavior halfway between the non-linear VFT behavior and the Arrhenius *high temperature* one, since under certain conditions of minimum moisture, the *high temperature* relaxation process is observed in the whole temperature range before the onset of thermal degradation as it is shown in Figures 5 and 6. For this *high-temperature* relaxation, non-neutralized chitosan films seem to be more sensitive to water, since in the *dry* state presents a particular behavior in the 70-100°C temperature range, which is recognized as a superposition of two relaxation processes, giving raise a different slope value as it is shown in window inset of Figure 5a.

As mentioned before, there exists a great controversy about the glass transition temperature (T_g) of chitosan; while some authors show no evidence of a glass transition by DSC, DMTA and dielectric measurements [11-14, 26, 28], others report a wide variety of values (see Table 1). Some authors assigned values very close to chemical degradation [7-9], others seem to correspond to water elimination rather than a glass transition [6]. Some repotted values that lie in the temperature range proposed in the present work [3, 22, 27]. The controversy and particularly the discrepancy in the glass transition temperature of chitosan may be related to an inefficient elimination of water, a heat treatment near the degradation temperature of chitosan, the film preparation technique or neutralization process. Regarding the deacetilation degree (%DD), our dielectric results showed no DD-effect on the T_g in agreement with the report of Dong et al. [5]. However, without doubt, the moisture content is a key factor that determines whether or not the glass transition temperature in chitosan can be observed since as it can be seen, the presence of water drastically affects the chitosan backbone mobility, especially in the α-relaxation region that corresponds to the cooperative motion and as a result, moisture content is probably the main cause of the wide glass transition temperatures range reported in literature.

High Temperature and Low Frequency Relaxation

The "*high temperature*" was defined in the temperature from 70°C to the onset of degradation $\approx$210°C (Figures 5 and 6). It is well described by the Arrhenius model and it was detected in both neutralized and non-neutralized chitosan. The slope of these curve represent the activation energy of each process. Both, the temperature dependence of *dc conductivity* as well as that for *relaxation time* are Arrhenius-type $\sigma = \sigma_0 \exp(-E_{a\sigma} / RT)$ and $\tau = \tau_0 \exp(E_{a\tau} / RT)$). For this relaxation, non-neutralized chitosan films seem to be more sensitive to water, since in the *dry* state presents a particular behavior in the 70-100°C temperature range, which is recognized as a superposition of two relaxation processes, giving raise a different slope value (see window insert of Figure 5). This relaxation has Havriliak and Negami fitting parameters α =0.72 $\pm$ 0.08 and β = 1.0, these parameters are temperature independent in both chitosan forms in agreement with previous studies [17]. It can be seen that the experimental data could be fitted with the model involving less parameters than the HN model in agreement with other authors [52, 53].

Our activation energy calculations ($E_{a\sigma}$ 80-88.2 kJ/mol for *conductivity* and $E_{a\tau}$=80-89 kJ/mol for *relaxation time*) are in agreement with previous reports for neutralized and non neutralized chitosan [17], and for polysaccharides [16]. The activation energy values for both films are quite close. This process so-called σ-relaxation has been widely studied and is associated with the hopping motion of ions in the disordered structure of the biomaterial [16]. Einfeldt et al. [16] observed this relaxation process in the high temperature range (>80°C), however, in our case, on minimum moisture conditions this relaxation process discloses in the whole temperature range until the onset of thermal degradation making difficult the α-relaxation detection. For the case of non-neutralized films, moisture has a strong effect on the α-relaxation. This is shown in the *dry* non-neutralized samples in the 70-100°C temperature range with moisture content in the range of 0.05-3.5 wt% (see window insert Figure 5a). Note that a lower slope (lower activation energy) ascribed to the superposition of α-and σ-relaxations describes this "new" Arrhenius-type relaxation for non-neutralized films. Activation energy is ca. $E_{a\sigma}$ (conductivity) ≈ 28.8 and $E_{a\tau}$ (time relaxation) ≈38.34 kJ/mol which is much smaller than that of neutralized films with similar moisture content (~80-85 KJ/mol,). The activation energy value of the relaxation time for the σ-relaxation process in the *dry* samples shows that water exerts a greater effect on the non-neutralized chitosan form; this is because of its superior ability to form hydrogen bonds providing a lower activation barrier for motion of ions. In this chitosan form, the σ-relaxation process is shifted to slightly higher frequencies compared to neutralized chitosan (not shown), this entails lower relaxation times (Figure 6). The ion mobility in non-neutralized chitosan is facilitated by NH_3^+ groups providing it higher conductivity.

Pizzoli et al.[11] by dynamic mechanical analysis and dielectric measurements observed a "high-temperature relaxation" near 140°C; the calculated activation energy was ~100kJ/mol, and this value is in agreement with the activation energy of the σ-relaxation mentioned before. They did not interpret it as a glass transition and suggested that this relaxation arises from a molecular motion having a less co-operative character than the glass-to-rubber transition. Some authors have attributed this relaxation to the glass transition [4, 5], nonetheless it seems to be ion motion that yields this peak in dynamic mechanical spectra and not the glass transition, because the activation energy for segmental mobility should be greater.

Finally, in the low frequency range, a change from positive slope in the conductivity (Figure 5) to negative is disclose at 210°C and above denoting the onset of degradation, at this temperature the dependence of *resistance* and *capacitance versus temperature* also experiment a change in the slope (Figure 2a). TGA (Figure 2b) and DSC (not shown) measurements confirm this degradation.

High Frequency Relaxation

The well known high-frequency β-relaxation is identified in the 10^4-10^8 Hz frequency region for *dry* and *dry annealed* neutralized and non-neutralized films, however, in *wet* (12.8 wt % of moisture content) samples this relaxation is not well resolved and model-fitting is complicated. This is the reason why we cannot identify the β_{wet} relaxation process documented by other authors ascribed to biopolymer-swollen water motion [11, 33]. For *wet* polysaccharides, Enfieldt, et al.[16] reported a superposition of two relaxation processes: β and β_{wet} relaxations that merge into one common β-relaxation process as water is driven off. The temperature dependence of the relaxation time for this high frequency relaxation is found to be Arrhenius-type (not shown) with an excellent linear fitting of the *relaxation time* as a function of *1000/T*. Our calculated activation energy values are in excellent agreement with previous reports [16, 17, 33]. This β-relaxation is a common reported relaxation process in chitosan [11, 16, 17, 22] and it has been related to side group motions by means of the glucosidic linkage [16, 54]. This is the main relaxation process found in all pure polysaccharides in the low temperature range (-135 to + 20°C) and corresponds to local chain dynamics [16]. For chitosan, we have found this secondary relaxation in the high temperature range in agreement with other authors (higher than 20°C) [16, 33], however, it is expected to appear in the high temperature range at the very high frequency end since Einfeldt et al.[16] found this high frequency relaxation at temperatures as high as 120°C.

DYNAMIC MECHANICAL ANALYSIS (DMA)

Effect of Moisture Content on DMA Relaxations

Neutralized and Non-neutralized chitosan films moisture weight-percent was determined by TGA measurements under dry air flow emulating the DMA measurements environment. Figure 7 shows a weight loss of 11% for *wet* samples, 5.5 % for *dry* ones and 1.2% for *dry annealed* films. Moisture content in *wet* samples under dry air flow conditions are in agreement with our measurements reported for dielectric analysis. Chitosan DMA analysis shows three clear events (see Figure 8a) in non neutralized-chitosan films under *wet* and *dry* conditions; they can be distinguished by the presence of three *tan δ* peaks at 74°C, 182°C and 290°C in *wet* samples and at 88°C, 182°C and 290°C in films *drying* at 150°C before DMA measurements. These three relaxations are observed in neutralized films as well, it is shown in Figure 8.

Results and Discussion 33

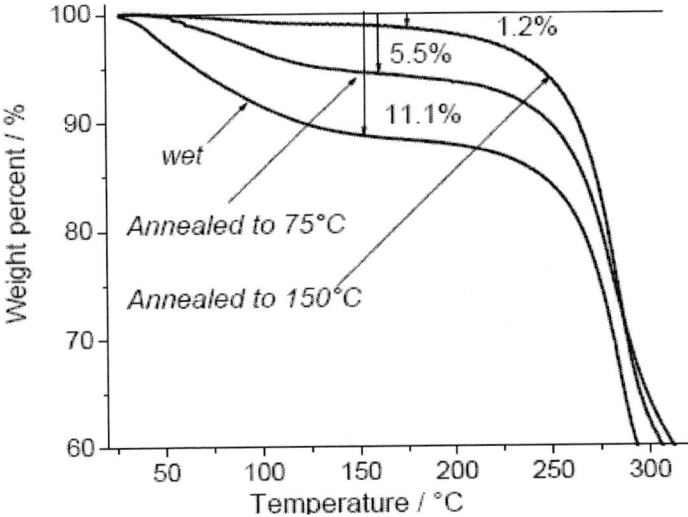

Figure 7. Thermogravimetric analysis for neutralized chitosan under air flow environment.

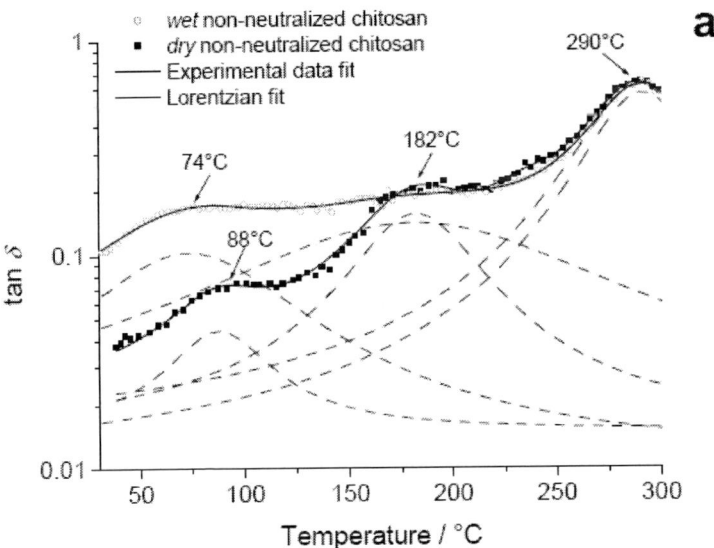

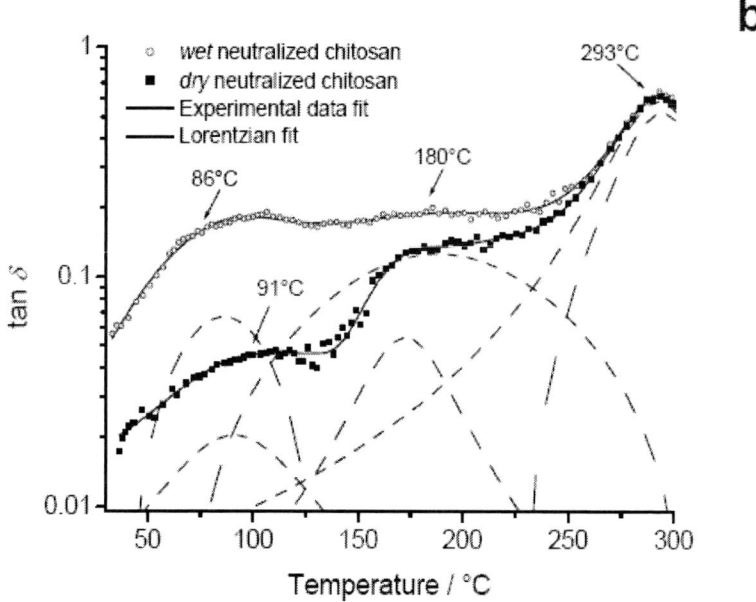

Figure 8. Viscoelastic properties for *wet* and *dry* (a) non-neutralized chitosan and (b) neutralized chitosan. Note, three main events are distinguished under *wet* conditions, storage and loss module occur at lower temperatures when comparing with tan δ peaks.

RELAXATION BELOW 100°C

The molecular relaxation occurring below 100°C is highly affected by moisture content; the higher the moisture content, the lower the relaxation temperature, i.e., for *wet* samples the relaxation temperature shifts to lower temperatures in contrast to *dry* samples (see Figures 8 a and b). This phenomenon can be related to a plasticizing effect of water on the glass transition temperature of chitosan. The plasticizing effect is more clearly observed in Figure 9 since for *wet* samples the damping tan δ peak is present at 54°C whereas for samples dried at 75°C and at 150°C it shifts to 81°C and 89°C respectively. However, as well as it was shown by impedance measurements, this relaxation vanishes in neutralized chitosan films after the second heat treatment as it is shown in window insert of Figure 9. Quijada-Garrido et al. [31] by means of DMTA measurements assigned this relaxation process occurring below 100°C as the α-relaxation and hence as the glass transition temperature of chitosan, however, they do not provide a theoretically sound explanation about this statement and do not give additional data related to the effect of moisture content on this relaxation. They set the glass transition temperature of chitosan at 85°C, and observed a decrease in the temperature of the maximum of the α-relaxation with increasing glycerol content, suggesting a plasticizing effect on chitosan.

Lazaridou et al. [21] showed the thermomechanical behavior of chitosan and its blends with starch or pullulan. The location of the peak that they denoted as the α-relaxation for pure chitosan shifts to lower temperatures with increase of water content or addition of sorbitol due to plasticization of the polymer matrix. The glass transition temperature varies from 77°C at 10% of water to -23°C at about 30% of water and they estimated a glass transition temperature using the Gordon-Taylor equation for *dry* chitosan about 94.9°C. M. Mucha et al.[4] showed two clear tan δ peaks occurring at 24°C and 43°C, moving to higher temperatures (32°C and 45°C respectively) in a second scan after heating at 130°C. Mucha et al. state that these peaks shift to higher temperatures because a chitosan film has lower amount of water, thus the system appears stiffer.

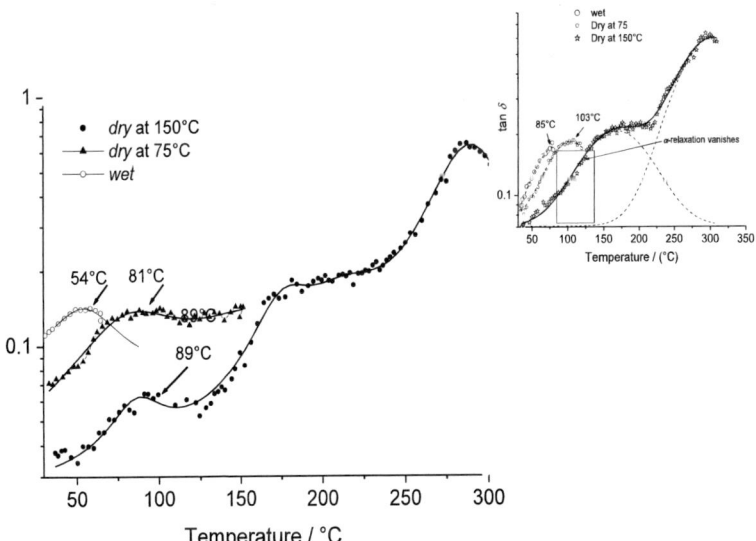

Figure 9. tan δ for non-neutralized chitosan. Note: the low temperature relaxation shifts to higher temperatures when moisture content decrease. Window insert: Neutralized chitosan, as well as for dielectric analysis, α-relaxation vanishes after annealing.

These two peaks were recognized as a structural reorganization of packing of chitosan molecules due to an increase of residual water mobility, volume expansion and following change of hydrogen bond strength.

Toffey et al.[27] reported a tan δ peak for wet samples ranging form 60°C to 93°C depending on the acidic media employed for chitosan dissolution, they used four acids: formic, acetic, propionic, and butyric acid and do not performed neutralization prior to DMA measurements. Likewise, in a different paper Toffey et al.[27] observed tan δ transitions for chitosanium acetate located at progressively higher temperatures, indicating a shift of T_g to higher temperatures as the temperature range over the sample scanned increases. This rising temperature is usually attributed to the removal of residual moisture which plasticizes the material; nonetheless, in the case of chitosanium acetate, surprisingly, Toffey et al. [27] claim that this effect is due to conversion of chitosanium acetate to chitin instead of loss of residual moisture.

RELAXATION ABOVE 100°C AND BELOW 200°C

The second tan δ peak located above 100°C (see Figures 8 and 9) had also been assigned as the α-relaxation of chitosan; Ogura et al.[23] by means of the loss peak proposed a glass transition temperature around 140°C, however, they do not report moisture content on chitosan films. Dong et al.[5] suggest a glass transition temperature for chitosan at 140°C-150°C, however, the DMA curves were obtained after the first run of heating up to 180°C and this temperature is high enough to induce thermal degradation in chitosan [39, 55], they also obtained a first peak at around 85°C ascribed to a water-induced relaxation. While M. Mucha et al.[4] denoted a tan δ peak occurring at 156-170°C as the α-relaxation at T_g of chitosan according to deacetylation degree, however they do not give a theoretical statement to support this issue. Likewise, Y.B.Wu et al.[29] designed the glass transition temperature of chitosan at 153°C by means of a tan δ peak.

RELAXATION ABOVE 200°C

Surprisingly, the third peak disclosed above 200°C, has also been assigned has the glass transition temperature of chitosan. Sakurai et al.[7] carried out DMA measurements in two cycles of heating: the first heating up to 180°C and the second heating up to 250°C after holding at 180°C for 5 min. In the first heating one large peak near 153°C is disclosed, however, the tan δ curve drastically changed in the second heating and a large shoulder appears at about 205°C, while the peak near 153°C disappears. They claim that the transition occurring at 205°C is related with the glass transition of chitosan, however, as it was shown above by TGA and dielectric measurements and supporting with literature reports [39], there is enough evidence to state that after the first heating up to 180°C chitosan is degradated. That is why the relaxation at 153°C disappears; therefore the process occurring at 205°C should be ascribed to the beginning of thermal degradation of the biopolymer and cannot be related with the glass-rubber transition.

DYNAMIC MECHANICAL ANALYSIS (DMA) AND DIELECTRIC SPECTROSCOPY MEASUREMENTS CONNECTION

As it was shown above, the understanding of dynamic mechanical analysis (DMA) experimental data and therefore the designation of the relaxation nature of each process disclosed in chitosan is a controversial issue. Without additional evidence it is difficult to identify the nature of each relaxation processes. In this case, DMA measurements support the dielectric spectroscopy results previously described.

RELAXATION BELOW 100°C

The "low frequency-low temperature" relaxation (20-80°C) obtained by dielectric spectroscopy as well as the relaxation disclosed below 100°C by DMA measurements is highly affected by moisture content, it suggests a plasticizing effect of water on the biopolymer. Dielectric results showed a non-Arrhenius behavior of this relaxation process that can be well described by the Vogel-Fulcher Tammann (VFT) equation, therefore, the first peak (dependent on moisture content) obtained below 100°C is related with the α-relaxation and consequently to the glass transition temperature of chitosan. The difference between the T_g value calculated by dielectric measurements and obtained by DMA measurements has been previously reported [31], for many polymers, mechanical relaxation processes were observed at different temperatures than the corresponding dielectric relaxation [31]. This in part because dielectric spectroscopy is sensitive to fluctuations of dipole moments and mechanical relaxation monitors the fluctuations of internal stresses. It is noteworthy that in dielectric as well as DMA measurements, this relaxation is detected not any more after the second annealing.

RELAXATION BETWEEN 100°C AND 200°C

The second peak occurring above 100°C and below 200°C, can be related to the σ-relaxation obtained by dielectric measurements. The activation energy disclosed by this relaxation process is in agreement with previous results reported for chitosan [16, 17], this process is associated with the hopping motion of ions in the disordered structure of the biomaterial.

RELAXATION ABOVE 200°C

A change in the slope by dielectric measurements at 240°C has shown the beginning of chitosan thermal degradation; this statement is supported by DSC and TGA measurements as well as literature previous reports [39]. Therefore, the third peak obtained above 200°C is related with thermal degradation of chitosan.

Chapter 8

CONCLUSIONS

The molecular dynamics of neutralized and non-neutralized chitosan was studied by dielectric spectroscopy and supported by dynamic mechanical analysis. The low frequency α-relaxation associated to the glass transition can be detected by dielectric spectroscopy once a pre-treatment of experimental data is performed. This relaxation process seems to be independent of the chitosan form evaluated (neutralized and non-neutralized) and is strongly influenced by moisture content, a glass transition can be calculated depending upon moisture content.

A plasticizing effect on chitosan α-relaxation is observed by dielectric spectroscopy and is supported by DMA analysis. For moisture contents less than 0.05 wt% the glass transition is difficult to observe due to a superposition of the α and σ-relaxation process. This work propose a successful method to monitor the plasticizing effect of water on chitosan glass transition by dielectric and DMA measurements, the same methodology is applied to TGA measurements in order to obtained the moisture weight percent with good agreement.

The well known σ-relaxation often associated with proton mobility is also observed in chitosan (neutralized and non-neutralized) from 80°C to the onset of degradation. On minimum moisture content conditions, this relaxation process could be detected in the whole temperature range before the onset of thermal degradation. It is strongly affected by moisture content for *dry* samples, by water effects, the activation energy shifts to lower values when compared to *dry annealed* samples. The non-neutralized chitosan showed an easier mobility in this ion motion process. This relaxation process exhibits a normal Arrhenius type temperature dependence with activation energy of 80-90 kJ/mol.

Finally, the high frequency secondary β-relaxation is also observed with Arrhenius activation energy of 46-48 kJ/mol.

REFERENCES

[1] Anderson, P.W. *Science,* 1995, 1615-1616.
[2] González-Campos, J. B.; Prokhorov, E.; Luna-Bárcenas, G.; Fonseca-García, A.; Sanchez. I. C. *J. Polym. Sci.* Part B:Polym Phys. 2009, 47, 2259-2271.
[3] Ratto J.; Hatakuyama T.; Blumstein R. B. *Polymer.* 1995, 427, 2915-2919.
[4] Mucha, M.; Pawlak, A. *Thermochim. Acta* 2005, 427, 69-76.
[5] Dong, Y.; Ruan, Y.; Wang, H.; Zhao, Y.; Danxia, B. *J. Appl. Sci.* 2004, 93, 1553-1558.
[6] Liu, Y.; Zhang, R., Zhang, J ; Zhou, W.; Li, S. *Iranian Polym. J.* 2006, 15, 935-942.
[7] Sakurai, K.; Maegawa, T.; Takahashi, T. *Polymer.* 2000, 41, 7051-7056.
[8] Rao, V.; Johns, J. *J. Therm. Anal. Cal.* 2008, 92, 801-806.
[9] Mayachiew, P.; Devahastin S. *Drying Technol.* 2008, 26, 176-185.
[10] Arvanitoyannis, I. S.; Kolokuris, I.; Nakayama, A.; Yamamoto, N.; Aiba, S. *Carbohydr. Polym.* 1997, 34, 9-19.
[11] Pizzoli, M.; Ceccorulli, G.; Scandola, M. Carbohydr Res 1991, 222, 205-213.
[12] Kittur, F. S.; Harish P.; K.V.; Udaya, S. K.; Tharanathan, R. N. *Carbohydr. Polym.* 2002, 49, 185-193.
[13] Mano, J. F. *Macromol. Biosci.* 2008, 8, 69-76.
[14] Arvanitoyannis, I. S.; Nakayama, A.; Aiba S. *Carbohydr. Polym.* 1998, 37, 371-382.
[15] Raju GG. Dielectrics in Electrical Fields. Marcel Dekker Inc.: New York, 2003. pp. 138, 157, 248, 259.

[16] Einfeldt, J.; Meiβner, D.; Kwasniewski, A. *Prog. Polym. Sci.* 2001, 26, 1419–72.
[17] Viciosa, M. T.; Dionisio, M.; Silva, R.M.; Mano, J.F. *Biomacromolecules.* 2004, 5, 2073-2078.
[18] González-Campos, J.B.; Prokhorov, E.; Luna-Bárcenas, G.; Mendoza-Galván, A.; Sanchez, I. C.; Nuño-Donlucas, S. M.; García-Gaitan, B.; Kovalenko Y. *J. Polym. Sci.* Part B:Polym Phys. 2009, 932-943.
[19] Pissis, P. Electromagnetic Aquametry; Water in Polymers and Biopolymers by Dielectric Techniques; Springer: Berlin Heidelberg, 2005.
[20] Neto, C. G. T.; Giacometti, J. A.; Job, A. E.; Ferreira, F. C.; Fonseca, J. L. C.; Pereira, M. R. *Carbohydr. Polym.* 2005, 62, 97-103.
[21] Lazaridou, A.; Biliaderis, C. G. *Carbohydr. Polym.* 2002, 48, 179-90.
[22] Kaymin, I. F.; Ozolinya, G. A.; Plisko, Y. A. *Polym. Science* 1980, 22, 171-177.
[23] Ogura, K.; Kanamoto, T.; Itch, M; Miyashiro, H.; Tanaka, K. *Polym. Bullentin.* 1980, 2, 301-304.
[24] Kim, S. S.; Kim, S. J.; Moon, Y. D.; Lee, Y. M. *Polymer.* 199,35, 3212-3216.
[25] Guan, Y.; Liu, X.; Zhang, Y.; Yao, K. *J. Appl. Polym. Sci.* 1998, 67, 1965-1972.
[26] Dufresne, A.; Cavaillé, J. Y.; Dupeyre, D.; Garcia-Ramirez, M.; Romero, *J. Polymer.* 1999, 40, 1657-1666.
[27] Toffey, A.; Glasser, W. G. *Cellulose.* 2001, 8, 35-47.
[28] Zohuriaan, M. J.; Shokrolahi, F. *Polym. Test.* 2004, 23, 575-579.
[29] Wu, Y. B.; Yu, S. H.; Mi, F. L.; Wu, C. W.; Shyu, S. S.; Peng, C. K.; Chao, A. C. *Carbohydr. Polym.* 2004, 57, 435-440.
[30] Shieh, Y. T.; Yang, Y. F. Euro Polym J. 2006, 42, 3162-3170.
[31] Quijada-Garrido, I.; Oglesias-González, V.; Mazón-Arechederra, J.M.; Barrales-Rienda, J.M. *Carbohydr. Polym.* 2007, 68, 173-186.
[32] Lee, Y. M.; Kim, S. H.; Kim, S. *J. Polymer.* 1996, 37, 5897-5905.
[33] Viciosa, M.T.; Dionisio, M.; Mano, J.F. *Biopolymers.* 2005, 81, 149-59.
[34] Dionísio, M.; Alves, N.M.; Mano, J.F. Review: Molecular dynamics in polymeric systems. e-polymers. 2004, 44. http://www.cqfb.fct.unl.pt/drs/docs/ mano_030704.pdf
[35] Runt, J. P.; Fitzgerald, J. J. Dielectric Spectroscopy of Polymeric Materials; fundamentals and applications. American Chemical Society, Washington, DC 1997. Pp 88, 89.

References

[36] Einfeldt, J.; Meißner, D.; Kwasniewski, A. *J. Non-Cryst Solids* 2003, 320, 40-55.
[37] Menard, K. P. Dynamic mechanical analysis: a practical introduction. CRC Press, 2nd edition. Boca Raton Fl, USA 2008. pp 2.
[38] Ahn, J. S.; Choi, H. K.; Cho, C. S. *Biomaterial.* 2001, 22, 923-928.
[39] Machado, A.A. S.; Martins, V. C. A.; Plepis, M.G. *J. Therm. Anal. Cal.* 2002, 67, 491-498.
[40] Yang, J. M.; Su, W. Y.; Leu, T. L. Yang, M. C. J Membr Sci. 2004, 236, 39-51.
[41] Ikejima, T.; Inoue, Y. *Carbohydr. Polym.* 2000, 41, 351-356.
[42] Mirzadeh, H.; Yaghobi, N.; Amanpour, S.; Ahmadi, H.; Mohagheghi, M. A.; Hormozi, F. *Iranian Polym. J.* 2002, 11, 63-68.
[43] Yanagawa, K. 4294A Precision Impedance Analyzer Operation Manual. *Agilent Technologies.* Japan 2000.
[44] Wintle, H. J. Conduction processes in polymers. In: Engineering dielectrics. Vol IIA, Electrical properties of solid insulating materials: Molecular structure and electrical behavior. Bartnikas, R., Eichorn, R. M., Eds.: ASTM Especial Technical Publication 783: Philadelphia, PA. 1983.
[45] F. Kremer, A. Schonhals (Eds.) Broadband Dielectric Spectroscopy, Springer-Verlag, Berlin 2003.
[46] Rowland, S.P Water in polymers. American Chemical Society, Washington D.C. 1980; Vol 1, pp.
[47] Stickel, F.; Fischer, E. W.; Richert, R. *J. Chem. Phys.* 1996,104,2043 2055.
[48] Cutroni, M.; Mandanici, A. *J. Chem. Phys.* 2001,114,7118-7123.
[49] Schönhals, A.; Kremer, F.; Hofmann, A.; Fischer, E.W. *Physical Review Letters.* 1993,70,3459-3462.
[50] Garcia, F.; Garcia-Bernabe, A.; Compan, V.; Diaz-Calleja, R.; Guzman, J.; Riande, E.; *J. Polym. Sci. B Polym. Phys.* 2001, 39, 286-299.
[51] Hodge, R.M.; Bastow,T.J. G.; Edward,H.; Simon, G.P.; Hill, A. *J. Macromolecules* 1996, 29, 8137-8143.
[52] Despond, S.; Espuche, E.; Cartier, N.; Domard, A. *J. Polym. Sci.* Part B Polym Phys 2005, 43, 48-58.
[53] Yagihara, S.; Yamada, M.; Asano, M.; Kanai, Y.; Shinyashiki, N.; Máximo, S.; Ngai, K. L. J. *Non-Cryst Solids* 1998, 235-237, 412-415.
[54] dos Santos Jr, D. S.; Goulet, P. J. G.; Pieczonka, N. P.W.; Oliveira Jr, O. L.; Aroca, R. F. *Langmuir* 2004, 20, 10273-10277

[55] Balau, L.; Lisa, G.; Popa, M. I.; Tura, V.; Melnig, V. *Cent. Eur. J. Chem.* 2004, 2, 638-647.

INDEX

A

acetic acid, 15
acid, 14, 15, 38
acrylate, 31
activation energy, vii, 28, 32, 33, 34, 40, 41, 42
amines, 30
amorphous phases, 6
amorphous polymers, 6, 7
amplitude, 13, 17
annealing, 16, 20, 21, 30, 37, 40
argon, 16
Arrhenius dependence, 28
atoms, 3
attribution, 11

B

behaviors, 3
bias, 18, 19
biopolymer, 2, 21, 34, 39
biotechnology, 2, 19
blends, 37
bonds, 30

C

calorimetric measurements, 6
cellulose, 5, 10, 28
character, 33
chemical degradation, 31
chemical properties, 2
chitin, 1, 2, 5, 11, 28, 38
clustering, 30
compensation, 17
composites, 16
compounds, 30
conditioning, 16
conductivity, 11, 21, 22, 25, 26, 27, 28, 29, 31, 32, 33
cooling, 16
copyright, iv
cristallinity, 3
crystalline, 3, 6
crystalline solids, 3
cycles, 38

D

damping, 36
decomposition, 14
degradation, 25, 32, 33, 40, 41
derivatives, 10
detection, 33
deviation, 18
dielectric constant, 8, 21
dielectrics, 46
differential scanning, 1, 17

Index

differential scanning calorimetry, 1, 17
dipole moments, 40
discontinuity, 18
dispersion, 7
distilled water, 15
DMA analysis, 2, 34, 41
donors, 30
drying, 5, 34
DSC, 1, 2, 4, 5, 13, 17, 31, 33, 40
dynamic mechanical analysis, 33, 39, 41
dynamics, 2, 7, 10, 11, 34, 45

E

electrical conductivity, 21
electrical properties, 21
electrodes, 15
elongation, 1
endothermic, 5
equilibrium, 17
evaporation, 20, 21
exposure, 3
extraction, 3

F

fabrication, 1
film thickness, 25
films, vii, 2, 3, 13, 15, 16, 18, 19, 20, 21, 23, 25, 26, 28, 29, 30, 31, 32, 34, 36, 38
fluctuations, 10, 40
food industry, 2, 19
free volume, 30
frequencies, vii, 18, 33
FTIR, 16

G

glass transition, vii, 1, 2, 3, 4, 5, 6, 7, 8, 10, 11, 13, 14, 19, 23, 26, 28, 29, 30, 31, 33, 36, 37, 38, 39, 41

glass transition temperature, vii, 1, 2, 3, 10, 14, 26, 30, 31, 36, 37, 38, 39
glasses, 3
glycerol, 37

H

heat capacity, 1, 5
heating rate, 17
hydrogen, 30, 33, 38
hydrogen bonds, 30, 33

I

ideal, 8
interface, 17, 18
ions, vii, 10, 32, 40

J

Japan, 46

K

KBr, 16

L

linear dependence, 28
liquids, 28, 29
low temperatures, 13

M

macromolecules, 2
manufacturing, 3
materials science, 2, 19
matrix, 21, 29
mechanical stress, 3
media, 38
melting, 5, 6

metals, 3
methodology, 41
micrometer, 16
migration, 2
modulus, 1, 13
moisture, vii, 13, 14, 19, 20, 21, 25, 26, 27, 29, 30, 31, 32, 34, 36, 37, 38, 39, 41
moisture content, vii, 13, 14, 19, 20, 21, 25, 26, 27, 29, 30, 31, 32, 33, 34, 36, 37, 38, 39, 41
molecular dynamics, 2, 7, 10, 41
molecular weight, 3, 15
molecules, 3, 14, 19, 21, 30, 38
Moon, 45

N

nitrogen, 17

O

organic polymers, 3
oscillations, 30
oxygen, 30

P

parameter, 3
permission, iv
permittivity, 8, 22
physical properties, 1, 13, 19
physics, 1
plasticization, 30, 31, 37
plasticizer, 19
polarization, 18, 21, 22, 25
polymer, 1, 3, 5, 6, 7, 10, 29, 30, 31, 37
polymer chains, 31
polymer matrix, 5, 37
polymer systems, 29
polymeric chains, 31
polymeric materials, 5, 7

polymers, 1, 2, 5, 13, 22, 29, 30, 39, 45, 46
potassium, 16
properties, vii, 1, 2, 3, 13, 19, 21, 26, 36, 46
protons, 28
purification, 15
purity, 16

R

relaxation, vii, 2, 4, 5, 6, 7, 8, 10, 11, 13, 14, 18, 21, 23, 25, 26, 27, 28, 29, 30, 31, 32, 33, 36, 37, 38, 39, 40, 41, 42
relaxation process, vii, 2, 7, 8, 10, 11, 23, 26, 28, 30, 31, 32, 34, 36, 39, 40, 41
relaxation processes, vii, 2, 7, 8, 10, 23, 31, 32, 34, 39
relaxation properties, 2
relaxation rate, 7
relaxation times, 7, 28, 29, 33
reliability, 3
resistance, 16, 18, 20, 22, 33
resolution, 16
respect, 10
retardation, 7
rights, iv
rings, 30
room temperature, 10, 21
rotations, 6
rubber, vii, 1, 5, 6, 13, 14, 33, 39
rubbers, 3
rubbery state, 5

S

semicircle, 18, 22, 25
semi-crystalline polymers, 1
shape, 7
shrimp, 25
solid state, 6
sorption, 30
space, 3

species, 25, 26
spectroscopy, vii, 2, 10, 16, 29, 39, 41
starch, 10, 37
stars, 20
stimulus, 7
storage, 36
synthesis, 1, 26

thermal expansion, 1, 5
thermal relaxation, vii, 13, 26
thermogravimetric analysis, 16
thin films, 2
transition temperature, 1, 14, 30, 31, 36, 37, 38

T

temperature, vii, 1, 2, 3, 4, 5, 7, 8, 10, 11, 13, 14, 17, 18, 20, 23, 25, 26, 27, 28, 29, 30, 31, 32, 33, 34, 36, 37, 38, 39, 41
temperature dependence, vii, 8, 10, 23, 28, 32, 34, 41
TGA, 16, 21, 26, 33, 34, 39, 40, 41
thermal activation, 8
thermal analysis, 2, 5, 13
thermal degradation, vii, 14, 25, 31, 33, 38, 39, 40, 41

V

vacuum, 10, 15, 16, 17, 21, 22, 30
vapor, 3
variations, 21
viscosity, 1, 7

W

water evaporation, 20, 26
water sorption, 30
weight loss, 16, 21, 34